Energy
106

阳光下的雪绒花

Edelweiss in the Sun

Gunter Pauli

[比] 冈特·鲍利 著

[哥伦] 凯瑟琳娜·巴赫 绘

高 芳 李原原 译

上海远东出版社

丛书编委会

主　任：田成川

副主任：何家振　闫世东　林　玉

委　员：李原原　翟致信　靳增江　史国鹏　梁雅丽
　　　　任泽林　陈　卫　薛　梅　王　岢　郑循如
　　　　彭　勇　王梦雨

特别感谢以下热心人士对童书工作的支持：

匡志强　宋小华　解　东　厉　云　李　婧　庞英元
李　阳　刘　丹　冯家宝　熊彩虹　罗淑怡　旷　婉
杨　荣　刘学振　何圣霖　廖清州　谭燕宁　王　征
李　杰　韦小宏　欧　亮　陈强林　陈　果　寿颖慧
罗　佳　傅　俊　白永喆　戴　虹

目录

Contents

ZERI Learning Initiative

在奥地利阿尔卑斯山上，一朵纯白色的雪绒花正享受着明媚的阳光，它的旁边生长着一株美丽的深蓝色的阿尔卑斯山龙胆草。

"我一直羡慕你毛绒绒的、白色的花瓣。"龙胆草说。

A pure white edelweiss is enjoying the bright sun in the Austrian Alps, where it is growing next to an Alpine gentian, with its beautiful deep blue colour. "I have always admired you for your plush, white petals," says the gentian.

在奥地利阿尔卑斯山上享受着明媚的阳光

Enjoying the bright sun in the Austrian Alps

为我写了一首优美的歌曲

Wrote a beautiful song about me

"谢谢。我的洁白让世界各地的许多人都成了我的朋友。我的奥地利同胞把我印到在欧洲各地流通的硬币上，我的美国朋友甚至为我写了一首优美的歌曲。"

"我不明白你为什么从不会被太阳晒伤。"龙胆草说。

"Thank you. My whiteness has made me many friends around the world. My fellow Austrians put me on a coin used all around Europe and my American friends even wrote a beautiful song about me."

"I have a hard time understanding how you never get sunburn," says the gentian.

"太阳根本无法影响我。"

"你厚厚的花瓣能阻挡阳光吗？"

"不能，我们根本阻挡不了任何东西，而是允许阳光穿透。"

"The sun doesn't affect me at all."

"Do your thick petals block the rays?"

"Not at all; we don't block anything. We let the sun come through."

你厚厚的花瓣能阻挡阳光吗?

Do your thick petals block the rays?

天上有一个大洞

A big hole in the sky

"但这不是很危险吗？我听说人类使用特殊的面霜来保护他们的皮肤。"

"人类的确应该保护他们自己，但也应该保护环境。这样天上就不会有一个让所有强烈的紫外线穿过的大洞了。"

"But isn't that dangerous? I hear people apply special creams to protect their skins."

"People should protect themselves, but they should also protect the environment. Then there wouldn't be a big hole in the sky that lets all these harsh sunrays through."

"天上有一个洞？

这会威胁成千上万人的健

康，你是怎样在这样的山脉中生存

下来的？"

"嗯，很难解释清楚，但是高山上的空

气比山谷里的更稀薄。当空气很稀薄

时，天气很干燥，阳光会更

容易穿过。"

"A hole in the sky?
How can you survive here in
the mountains with something
like that, which puts the health of
millions in danger?"
"Well, it's difficult to explain, but here
in the mountains, when you go up very
high, there is less air than down in
the valley. When the air is thin, it
is dry and easier for the sun to
get through."

空气比山谷里的更稀薄

Less air than down in the valley

我只是让它进来

I simply let it come in

"但你还没有告诉我全部。你怎么从来不会被晒伤？你必须告诉我你的秘密。"龙胆草强调。

"没什么秘密。我并没有隔离阳光，只是让它进来。"

"然后呢？"

"But you are not telling me the full story. How come you never get surnburnt? You have to tell me your secret," insists the gentian.

"It's no secret. Instead of keeping the sun out, I simply let it come in."

"And then?"

"光被分散成微小的射线，这样当它到达我的身体组织时，就不会对我造成伤害了。"

"这就和武术一样！你用敌人的力量对付他自己。"

"The light is scattered into tiny little rays, so that by the time it gets to my tissues, it's all gone."

"That's what you do in martial art! You use your enemy's energy against himself."

你用敌人的力量对付他自己

Use your enemy's energy against himself

我花瓣上的纤细绒毛

Tiny hairs on my petals

"有点像，
但不全是那样。不
是隔离我们的朋友——太阳，
而是让阳光进入我的身体，然后我
把能量输送到花瓣上成千上万的纤细绒
毛上。"

"真聪明！这比人类为了保护自己免受太
阳光照射而在窗户、地毯、汽车油漆甚
至皮肤上使用的有毒物质污染少
多了。"龙胆草说。

"Kind of, but
not really. Instead of
keeping our friend, the sun,
out, I let the sun in, but I channel
the power to thousands of tiny hairs
on my petals."

"Smart, indeed! It is so much less
polluting than the toxins humans
use on windows, in carpets, for car
paints, and even on their skin to
protect themselves from the
sun," says the gentian.

"但我认为
你应该分享更多藏在
你那些美丽的蓝色花朵下面的
小秘密……"
"好吧，我的秘密在我的根里。任何人
的胃有问题时，告诉我就行了。"龙胆
草眨眨眼说。

……这仅仅是开始！……

"But I think you
should be sharing more of
the little secrets you're hiding
underneath those beautiful blue
flowers of yours …"
"Well, I keep them guarded in my
roots, but whenever anyone has a
stomach problem, just let me know,"
says the gentian, and winks.

… AND IT HAS ONLY JUST
BEGUN!...

AND IT HAS ONLY JUST BEGUN!

Did You Know?

你知道吗？

The edelweiss originated in Asia and settled in Europe during the glacial periods. In Europe, it grows in the Jura, the Pyrenees, the Alps, and the Carpathians.

雪绒花源于亚洲，冰河时期来到欧洲。欧洲的侏罗山、比利牛斯山、阿尔卑斯山和喀尔巴阡山都有雪绒花。

"Edelweiss" is German for "noble white". Its Latin name, *Leontopodium*, refers to how the shape and the woolly appearance of the flower makes it looks like a lion's paw. There are 30 different kinds of edelweiss.

"雪绒花"在德语中的意思是"高贵的白色"。拉丁名是 *Leontopodium*，从名字可以看出，雪绒花的形状和毛绒绒的外观使它看起来像狮子的爪子。雪绒花有 30 个不同的品种。

The edelweiss, a member of the sunflower family, is a perennial flower that lives on limestone rocks up to 3 000 metres high. Its dense hair protects it against the cold, drought, and ultraviolet radiation.

雪绒花是向日葵家族中的一员，它是一种多年生草本开花植物，生长在高达 3 000 米的石灰岩上。浓密的绒毛能帮助它抵御寒冷、干旱和紫外线的照射。

The edelweiss is the national symbol of three countries: Austria, Switzerland, and Romania. Switzerland started to protect the flower in 1878 and was later joined by European and Asian nations like Iran, India, Indonesia, Malaysia, and Mongolia.

雪绒花是 3 个国家的国家象征：奥地利、瑞士和罗马尼亚。瑞士从 1878 年开始保护雪绒花，后来欧洲和亚洲的伊朗、印度、印度尼西亚、马来西亚和蒙古等国也加入进来。

雪绒花的根可以抵御强大的山风，它的绒毛形成浓密的毯子防止其汁液受冻。

The roots of the edelweiss can resist strong mountain winds and their dense blanket of tiny hairs prevents their sap from freezing.

积雪覆盖的减少以及岩石和黑土对太阳能的吸收，导致山地植被无法生长在其他地方并危及雪绒花的生存。阿尔卑斯山阻止了南部的植被不断向北生长，这会导致雪绒花的灭绝。

The decrease of the snow cover and the absorption of solar energy by rocks and dark soils have led to mountain vegetation not becoming established elsewhere and endanger the survival of the edelweiss. The Alps block southern vegetation from growing further north, leading to their extinction.

The song *Edelweiss* is a show tune of the musical *The Sound of Music*. This was first as a Broadway musical production in 1959 and then in 1965 the film adaptation followed. The edelweiss flower in the song was used as a symbol for loyalty to Austria. This popular tune is often thought to be an Austrian folk song, even though it is a Broadway composition.

《雪绒花》是音乐剧《音乐之声》中的一首歌。1959年，《音乐之声》作为一部百老汇音乐剧出品，并紧接着在1965年被改编成电影。歌里的雪绒花在奥地利是忠诚的象征。这首流行歌曲常常被认为是奥地利民歌，尽管它其实是一首百老汇歌曲。

The gentian grows worldwide and has been used for centuries as an herbal remedy to treat indigestion, hypertension, parasitic worms, and even malaria. It is a core ingredient of Bach flower remedies, a collection of natural medicine.

龙胆草生长于世界各地，被用作草药已经有几个世纪了，它被用来治疗消化不良、高血压、寄生虫，甚至疟疾等疾病。它是巴赫花精疗法（一种天然药物的组合）的核心成分。

Do you like the idea that a powerful force can become harmless because its strength dissipates into thousands of little forces?

当一股强大的力量被分散成成千上万的小股力量时，它会变得无害，你喜欢这个想法吗？

当你不得不抵御寒冷、大风和刺目的阳光，而成千上万人仍赞叹你的美丽时，你感觉如何？

How would you feel if you had to defy the cold, wind and harsh sun, and are then celebrated for your beauty by millions around you?

With whom would you be prepared to share your secrets?

你愿意向谁分享你的秘密？

你更愿意使用防晒霜、遮阳伞，还是某种类似雪绒花所采用的方式来保护自己免受太阳辐射？

How do you prefer to protect yourself against the sun: with sunscreen, a parasol, or protection like the one the edelweiss developed?

Do It Yourself!

How do you protect yourself against the sun? What kinds of sunscreens are popular? While we must protect ourselves against excessive exposure to the sun, we also need to expose ourselves regularly to the sun in order to produce vitamin D. We need to find a balance between avoiding and welcoming the sun on our skin. That balance is a dynamic one as not getting enough sun could cause your bones to become brittle. So, ask your friends and relatives who of them are protecting their skins all the time? And tell them that by doing that, they may be reducing their exposure to the sun, which is necessary to help absorb enough calcium into their bones.

如何保护自己免受太阳照射？什么样的防晒霜最流行？我们必须避免过度暴露在阳光下，同时，为了生成维生素 D，我们也需要经常晒晒太阳。在晒和不晒之间，我们需要找到一个平衡。这个平衡是动态的，因为没有得到足够的阳光照射可能会导致你的骨骼变得脆弱。所以，问问你的朋友和亲人，他们每时每刻都在防晒吗？告诉他们这样做可能会让他们更少暴露在阳光下，而晒太阳对于帮助骨骼吸收足够的钙是很有必要的。

学科知识
Academic Knowledge

生物学	雪绒花是一种多年生草本植物；雪绒花很慷慨，它供养昆虫纲下的29个科，但只有少数昆虫为其授粉；一株授粉的雪绒花每个季度只生成一颗种子，成熟期是20—30天；龙胆草有400种，是两年生或多年生常绿植物。
化 学	雪绒花生长在碱性环境中；雪绒花与昆虫共生，因为来自植物花蜜的氨基酸对于为其授粉的昆虫的生存是不可或缺的；龙胆草生长在中性偏酸性的土壤和岩石园中；氧化锌和氧化钛是最受欢迎的防晒成分；金属氧化物混合成聚合物，生产饮料包装、食品包装和化妆品。
物 理	雪绒花根状茎呈圆柱形；除了波长小于0.18微米的紫外线，大部分光都会被雪绒花的白色绒毛层反射掉，雪绒花的绒毛吸收紫外线，防止它穿透植物的"皮肤"；雪绒花的绒毛是透明的，但它看起来是白色的，就像北极熊，这是因为绒毛之间的空隙会散射光；臭氧吸收太阳紫外线辐射；悬浮在空气中的水粒子会阻挡太阳；越高的地方空气越稀薄，更多的阳光可以穿透大气层；紫外线会杀死微生物；鸟类、蜜蜂、爬行动物可以看到紫外线，紫外线可以改善它们的视力；紫外线能帮助昆虫导航。
工程学	大自然有时不是抵御而是顺应自然力量。
经济学	奥地利面值2分欧元的硬币上有雪绒花图案；防止紫外线和使用紫外线都是数十亿美元的商机；世界各地都在保护雪绒花，现在它被种植以用作化妆品原料。
历 史	《雪绒花》这首歌出自1959年美国百老汇的歌舞剧《音乐之声》，今天许多奥地利人认为这是一首古老的民歌。
地 理	阿尔卑斯山、喀尔巴阡山、侏罗山和比利牛斯山都在欧洲；臭氧层位于距地面30—50千米高的平流层。
生活方式	我们已经意识到要保护自己免受太阳过度照射，但是我们却忽视了我们也需要把自己暴露在太阳下。
社会学	德国军队组成的阿尔卑斯部队使用雪绒花作为制服的徽章，但在第二次世界大战结束时，雪绒花已经成为德国人民抵抗纳粹主义的象征。
心理学	人们视雪绒花这种植物为勇气和高贵纯洁的象征，因为它在严酷的气候下仍能生长，而且它拥有纯白色的外表。
系统论	雪绒花象征即使在恶劣的条件下也能生活，它美丽且激励人心；雪绒花作为生物医学产品的价值已被认可，而且它也是化妆品的一个关键成分。

情感智慧
Emotional Intelligence

雪绒花　　　　雪绒花在整个对话中都是有自知之明和自信的。对于奥地利赋予它的荣誉以及在百老汇音乐剧中起重要作用，她感到很荣幸。她的话语风格简明扼要。她不自夸，也不说太多自己的细节。雪绒花把重点从自己身上转移到环境上，表现出对生态系统的理解。雪绒花准备分享她如何保护自己免受紫外线过度照射。她提供了一些细节，同时保持语言简洁，展示了很好的沟通能力。当雪绒花揭示了她抵御阳光的秘密，龙胆草也只能向她解释自己藏在根里的秘密。

龙胆草　　　　龙胆草在对话的开始就表示了他对雪绒花的钦佩和共鸣。龙胆草希望了解雪绒花如何保护自己，并以极大的信心和坚持直接提出问题。当雪绒花把话题转向环境时，龙胆草用一句简单的请求把话题拉了回来："告诉我你的秘密。"当雪绒花给出说明后，龙胆草迅速跟进使问题更加清晰。当雪绒花表示他用武术来类比并不正确时，龙胆草不仅没有生气，反而为雪绒花巧妙的解决方案感到高兴。龙胆草愿意回报雪绒花的分享，解释说，他的主要优势在根上，它含有一种能用于治疗消化疾病的很好的药。

艺术
The Arts

找到《雪绒花》这首歌，找几个朋友和你一起在唱诗班唱这首歌。你可以清唱，就是说没有音乐伴奏地唱。这是一首短小而浪漫的歌曲。然而，不能只是唱这首歌，还要讨论在第二次世界大战背景下这首歌的幕后故事。你甚至可以花点时间看看《音乐之声》这部电影。

思维拓展
Systems: Making the Connections

在最困难的环境下也有生命。雪绒花接受了潜在的威胁生命的生存条件，通过进化以一种杰出的适应能力把这些条件变成了一个赞美她的美丽的机会。它生活在海拔3 000米的地方，那里常年有风，冬天天气寒冷，夏天空气稀薄干燥、氧气有限、阳光猛烈。雪绒花证明了在极端的条件下也会产生极致的创新。雪绒花的茎和装饰花瓣的成千上万的绒毛分散了紫外线，紫外线的影响在其到达植物的"皮肤"之前已经消散。一些潜在的伤害（紫外线）已被转化为中性的力量，这种力可能有温和的细菌控制效果。此外，雪绒花给近30个科的昆虫提供营养，但只有几种昆虫给它授粉作为回报。如此慷慨在自然界里是很少见的。然而，这些少数授粉昆虫依赖雪绒花获取生存所需的氨基酸，这就是共生和进化。

雪绒花起源于亚洲，后定居在欧洲各大山脉，激励年轻的登山者攀爬岩石去寻找雪绒花，在19世纪这种行为成为一种表达爱的流行方式，一些人勇于登峰并最终摘取雪绒花送给他们的情侣。在德国吞并奥地利之后，雪绒花这个纯爱和奉献的符号成为了抵抗纳粹、忠于国家的象征。这种花至今仍然继续激励着人们。雪绒花让我们深刻地理解，在一个复杂的环境下，技术、自然和社会行为是如何结合的。这种花富含象征意义，代表承诺、毅力和美德。美既在艺术中，也在科学中。

动手能力
Capacity to Implement

龙胆花很容易获得。去买一些根部完整的龙胆花。确保它的花没有暴露于恶劣的化学杀虫剂和杀真菌剂中。拿一块它的根，洗净后啃一啃。不要感到惊讶：它很苦！找出使用这个健康产品的最好方式。记住，它刺激消化，是促进消化的药物的主要成分。不用它来做饭，而是酿造一种健康的饮料，将它尽可能地与其他成分融合。确保你在家长的监督下进行尝试，在交给别人尝试前先检查一下。

故事灵感来自
This Fable Is Inspired by

让-波尔·维涅龙
Jean-Pol Vigneron

让-波尔·维涅龙（1950—2013）曾在那慕尔大学和列日大学学习物理学，并探索超越旧的科学学科定义的新科学前沿。他先研究半导体，后来进入美国IBM进行研究生学习。他的兴趣从计算机技术发展到光学和对各种透明材料的研究。在他与安德鲁·帕克（本丛书中的《不用画的色彩》就是受他的启发）的合作中，他发现了结构性色彩，这引导他开始对雪绒花进行研究。他是那慕尔圣母和平学院的教授，他教导学生：科学需要严谨的态度和训练，只有当我们探索梦想和欢迎不确定性时，科学中的新领域才会被发现。

图书在版编目（CIP）数据

冈特生态童书.第三辑修订版：全36册：汉英对照 /
（比）冈特·鲍利著；（哥伦）凯瑟琳娜·巴赫绘；
何家振等译.—上海：上海远东出版社，2022
书名原文：Gunter's Fables
ISBN 978-7-5476-1850-9

Ⅰ.①冈…Ⅱ.①冈…②凯…③何…Ⅲ.①生态环
境－环境保护－儿童读物—汉、英 Ⅳ.①X171.1-49

中国版本图书馆CIP数据核字（2022）第163904号
著作权合同登记号图字09-2022-0637号

策　　划　张　蓉
责任编辑　祁东城
封面设计　魏　来　李　廉

冈特生态童书

阳光下的雪绒花

［比］冈特·鲍利　著
［哥伦］凯瑟琳娜·巴赫　绘

高　芳　李原原　译

记得要和身边的小朋友分享环保知识哦！
八喜冰淇淋祝你成为环保小使者！